Llwyn Celyn

*The story of a late medieval
house in the Black Mountains*

*Hanes tŷ diwedd yr oesoedd
canol ym Mynydd Du*

The Landmark Trust

The Landmark Trust

Adferwyd Llwyn Celyn a'i glwstwr o dai allan yn 2018 gan y Landmark Trust.

Elusen gadwraeth adeiladau yw'r Landmark Trust; mae'n achub adeiladau hanesyddol sydd mewn perygl, cyn eu gosod ar gyfer gwyliau gan roi iddynt ddyfodol newydd. Caiff unrhywun aros yn Llwyn Celyn, neu yn unrhyw adeilad Landmark arall, gydol y flwyddyn.

Mae'r Ysgubor Ddyrnu yn Llwyn Celyn ar gael i'w rhentu ar wahân ar gyfer gweithgareddau cymunedol ac addysgol. Mae ystafell hysbysrwydd ar agor hefyd yn ystod oriau golau dydd. Ar ddyddiau penodol bob blwyddyn mae'r safle cyfan yn agored i'r cyhoedd yn rhad ac am ddim.

I ddysgu rhagor neu i drefnu'ch gwyliau yn Llwyn Celyn neu unrhyw adeilad Landmark arall, ewch at www.landmarktrust.org.uk neu ffoniwch Landmark Booking Enquiries ar 01628 825925.

Llwyn Celyn and its group of outbuildings were restored in 2018 by the Landmark Trust.

The Landmark Trust is a building preservation charity that rescues historic buildings at risk and gives them a new future by offering them for holidays. Anyone can stay at Llwyn Celyn, or any other Landmark building, throughout the year.

At Llwyn Celyn, the Threshing Barn is available to hire separately as a community and educational facility. There is also an information room open during daylight hours. The whole site is open free to the general public on certain days each year.

For more information or to book a holiday at Llwyn Celyn or any other Landmark, please visit www.landmarktrust.org.uk or call Landmark Booking Enquiries on 01628 825925.

Cynnwys
Contents

Rhagair
Introduction

Fel un o'r tai preswyl pwysicaf sydd wedi goroesi er yr Oesoedd Canol, mae Llwyn Celyn wedi'i restru fel adeilad Gradd 1. Mae gwaith saer godidog i'w weld ynddo, ynghyd â heulfa ddeulawr anghyffredin a godwyd ar yr un adeg â'i neuadd agored. Tua diwedd yr 17eg ganrif, ychwanegwyd llawr, simnai a grisiau yn y neuadd. Gellir olrhain datblygiad pensaernïaeth ddomestig drwy hanes y tŷ hwn.

Pan ddaeth Llwyn Celyn i ddwylo'r Landmark Trust yn 2008, roedd ar fin mynd â'i ben iddo. Adeiladwyd tŷ newydd gerllaw ar gyfer y ffermwyr a oedd wedi byw ynddo gynt. Erbyn 2014, roedd Landmark wedi llwyddo hefyd i godi'r cyllid a oedd ei angen er mwyn adfer Llwyn Celyn, gan gynnwys grant galluogi oddi wrth Gronfa Dreftadaeth y Loteri (HLF). Dechreuodd y gwaith adfer yn yr haf 2015; diolch i'r grant HLF, bu'n bosibl trefnu hyfforddiant i grefftwyr, gweithgareddau cymunedol, artistiaid preswyl a dyddiau astudio yn ogystal. Datguddiwyd llu o fanylion cyffrous, gan gynnwys tystiolaeth i Lwyn Celyn gael ei adeiladu ym 1420, dyddiad eithriadol gynnar sy'n troi'r ffaith iddo oroesi gyhyd yn wyrth.

Listed Grade I, Llwyn Celyn (meaning holly bush or grove) was long known to be one of the most important surviving domestic houses in Wales, with fine joinery and a rare two-storey solar cross-range. We now know it was built in 1420. In the late 17th century, the hall was ceiled over and a staircase and chimneystack were inserted. Little had changed since then, so that this evolution of domestic architecture can still be clearly read in the house.

Before restoration, Llwyn Celyn was in a perilous condition.
Cyn iddo gael ei adfer, roedd Llwyn Celyn mewn cyflwr enbydus.

Although still partially inhabited by its farmers, in 2008 Llwyn Celyn was in a desperate state. Emergency scaffolding had been in place since the early 1990s, the roof was leaking and water ran off the hillside through some of the ground floor rooms. Cadw and the National Heritage Memorial Fund offered Landmark grants to allow the purchase the house, its outbuildings and the field below. There followed a complicated acquisition process that included the construction of a new farmhouse for the owners, so that they could continue to farm their land around.

This acquisition process finally came to fruition in 2014, when Landmark also reached its appeal target. The funds included an enabling grant from the Heritage Lottery Fund (HLF), alongside generous donations of all sizes from Landmark's supporters and other private trusts. Work on site began in summer 2015; the HLF grant also enabled craft training and community activities, artists-in-residence and study days during the works.

Hanes Cryno Llwyn Celyn
Brief History of Llwyn Celyn

Hanes cynnar Priordy Llanddewi Nant Hodni

Mae cyswllt anorfod rhwng hanes Llwyn Celyn a Phriordy Llanddewi Nant Hodni, am iddo gael ei adeiladu ar diroedd y Priordy. Wedi Diddymiad y Mynachlogydd, arhosodd Llwyn Celyn yn rhan o Ystad Llanthony fel y'i gelwir yn Saesneg nes iddi gael ei chwalu tua diwedd y 1950au.

Sefydlwyd Priordy Awstinaidd Llanddewi Nant Hodni ym 1118. Mae'n debyg bod safle Llwyn Celyn ymhlith ei waddoliadau cynharaf. Roedd y priordy'n dân ar groen y trigolion lleol yn y dyddiau cynnar, ac yn ystod cyfnod yr aflonydd gwleidyddol ym 1135 penderfynodd y mynaich ffoi i Henffordd, ac yna erbyn 1137 i Gaerloyw lle sefydlasant chwaer-briordy Llanthony Secunda. Erbyn y 1170au, gan fod yr ardal wedi tawelu, dychwelodd y canoniaid i Landdewi Nant Hodni. Wedi i nawddogaeth newydd ymddangos, bu'n bosibl rhwng 1180 a 1230 iddynt godi'r eglwys a'r adeiladau eraill yn Llanthony Prima y mae eu hadfeilion i'w gweld o hyd heddiw.

Wedi i'r Pla Du dorri allan ym 1349 gan ddifrodi poblogaeth y wlad, bu prinder pobl i weithio'r tir. Crisialodd anniddigrwydd y Cymry ynghylch rheolaeth Lloegr o gwmpas Owain Glyn Dŵr, a wrthryfelodd ym 1400 yn erbyn Harri IV. Ym 1404-5 roedd Dyffryn Ewias, sef bro Llanddewi Nant Hodni, wrth galon yr aflonyddwch. Ysbeiliodd lluoedd Glyn Dŵr y priordy, a ffoi a wnaeth y rhan fwyaf o'r canoniaid drachefn i Henffordd neu briordy Llanthony Secunda.

Er i'r mynaich ddychwelyd tua 1415, ychydig a wyddys am hanes Llanthony Prima dros yr hanner can mlynedd nesaf, ac eithrio bod y mynaich wedi erfyn

The ruins of Llanthony Priory at the northern end of the Honddu Valley speak of its past wealth.
Mae adfeilion Priordy Llanddewi ar ben gogleddol Dyffryn Honddu yn tystio i'r golud a fu.

am gael eu heithrio rhag talu trethi, gan ddatgan bod 'eu tiroedd a'u heiddio wedi'u diffeithio i'r fath raddau gan y rhyfeloedd ....nes iddynt fethu â fforddio cynnal y gwasanaeth dwyfol na thalu am y clerigwyr cysylltiol'.

Serch hynny, adeiladwyd Llwyn Celyn ym 1420, flynyddoedd prin ar ôl i ddilynwyr Glyn Dŵr ddifetha'r priordy. Nid yw'n hysbys inni pwy adeiladodd y fath dŷ mawr ei fri, nac ychwaith sut defnyddid Llwyn Celyn cyn Diddymiad y Mynachlogydd. Erbyn 1481 roedd Priordy Llanddewi Nant Hodni wedi dirywio cymaint nes iddo gael ei gymryd drosto gan ei chwaerdy Llanthony Secunda yng Nghaerloyw. Syrthiodd Llwyn Celyn, ynghyd â gweddill eiddo Llanthony Prima, i ddwylo Llanthony Secunda, nes i'r ddau Briordy gael eu diddymu gan Harri VIII ym 1538.

The early history of Llanthony Priory

Llwyn Celyn's history is inextricably linked with Llanthony Priory, on whose lands it was built within the wider Vale of Ewyas. The extensive ruins of the priory have fascinated and inspired visitors for centuries since the priory's

Dissolution in 1538. Llwyn Celyn remained part of the Llanthony Estate until the estate was finally broken up in the late 1950s.

Around 1100, William de Lacy came upon a ruined chapel dedicated to St. David in the forest. William was a knight and kinsman of Hugh de Lacy, a powerful Marcher lord whose family were granted the Vale of Ewyas by William the Conqueror. Inspired by the peace and holiness of the spot, William dedicated himself to solitary prayer. The community grew, and a church dedicated to St John the Baptist was consecrated in 1108. In 1118, the Augustinian rule was adopted and the priory formally constituted. The little chapel's full name – Llandewi-nant-Honddu or St David's chapel by the Honddu river – became contracted to Llanthony.

Hugh de Lacy gave lands around the priory site, which became the home manor of Cwmyoy, in whose parish Llwyn Celyn still lies today. It seems likely that the Llwyn Celyn site was part of these earliest endowments.

The priory formed part of the wider policy of the Norman Lords to colonise and control Welsh lands, and the local population resented the monastic interlopers. Hostilities came to a head in 1135 during the Welsh uprising after the death of Henry I. The monks fled to seek refuge, first at the Bishop's Palace in Hereford. By 1137, they had established an alternative site at Gloucester, known as Llanthony Secunda (or Llanthony-by-Gloucester) to distinguish it from its mother house.

By the 1170s, peace had returned and the monks returned to Llanthony Prima. The de Lacy family gave more lands, including significant endowments in Ireland which enabled the building of the church and

Arms of Llanthony Priory found at Bromsberrow, on which the Landmark curtain design is based. Arfbais Priordy Llanddewi a ddarganfuwyd yn Bromsberrow.

claustral buildings at Llanthony Prima from *c.* 1180 to 1230, whose impressive ruins survive today.

Llanthony Prima was at the height of its wealth and power, but life was rarely easy in these medieval borderlands. The Poll Tax return of 1381 reveals only seven registered canons: the Black Death arrived in Wales in 1349 and was followed by successive waves of plague, decimating the population and disrupting land cultivation. The English response was to exert ever-greater pressure on the local population. Welsh discontent found its figurehead in Owain Glyn Dŵr, who in 1400 rebelled against Henry IV's rule and declared himself Prince of Wales.

It was a period of enormous upheaval and destruction as both armies raged to and fro. Abergavenny was burnt in 1404, although its castle held out. After defeat at Mynydd Cwndu, the Welsh forces retreated down the Usk Valley but ambushed the pursuing English near Mitchel Troy, killing many. There were English victories at

Glyn Dwr's seal. Sêl Owain Glyn Dŵr.

Grosmont and Usk in 1405. The Llanthony Valley was at the heart of this unrest. At some point in this chaotic year, Glyn Dŵr's forces sacked Llanthony Priory and most of the monks again fled to Hereford or Llanthony Secunda.

Glyn Dŵr's uprising was eventually suppressed by the superior English resources. Glyn Dŵr himself disappeared and his death was recorded by a former follower 1415. Very little is known of Llanthony Priory for the next fifty years or so, other than pleas for exemption from taxation, the monks 'having declared that their lands and possessions are so wasted by the wars of Henry IV in Wales, by deceases of their tenants and other mishaps and by impositions and charges, that they cannot support divine service and other incumbent charges ' (Cal. Close Rolls, 1 Jan 1448).

And yet, as we now know from timber analysis, Llwyn Celyn was built in 1420, just a few years after this destructive rising that ravaged the priory.

*Right: Llwyn Celyn
after restoration.
Dde: Llwyn Celyn
ar ddiwedd y gwaith
adfer.*

*Opposite: Llanthony
Priory, on whose
lands Llwyn Celyn
was built. Cyferbyn:
Priordy Llanddewi
Nant Hodni, y
codwyd Llwyn Celyn
ar ei diroedd.*

No documentary evidence survives from the time, and it is a puzzle who had
the resources to build such a high status house. The most likely explanation is
that Llwyn Celyn was initially intended as a house for the prior of Llanthony,
or possibly a steward for the Cwmyoy manor. Yet the priory was struggling
in these years. There is no evidence Llwyn Celyn had a private chapel such
as a prior would have required in his house (there is a 15th-century chapel
across the fields at Upper Stanton Farm but no direct connection has yet been
established). Stanton is said to have had its own manor house by the river,
although this is long gone.

Nor is anything known about how Llwyn Celyn was used or by whom for
the next century or so. Llanthony Prima was in decline, while its daughter
house in Gloucester, closer to sources of patronage and power, thrived. In
1481 Llanthony-by-Gloucester's powerful prior Henry Deane obtained from
Edward IV the union and subordination of Llanthony Priory to Llanthony
Secunda in Gloucester, on the grounds that the former had sunk so low
in liturgical observance and financial management. The documents are
frustratingly silent on Llwyn Celyn, but it passed with the rest of the Prima
holdings into the control of Llanthony Secunda.

In 1538, both Llanthony Priories were dissolved by Henry VIII, Llanthony-by
Gloucester's holdings worth considerably more than its rundown Welsh outpost.

Ar ôl y Diddymiad

Ym 1546 prynodd y cyfreithiwr cefnog Nicholas Arnold ystad Llanthony Prima, gan gynnwys Llwyn Celyn. Arhosodd yr ystad yn eiddo i'r teulu Arnold tan 1726, pan gafodd ei gwerthu i Edward Harley. Prynodd y Cyrnol Syr Mark Wood yr ystad ym 1799, a'i gwerthu eto ym 1807 i'r bardd Walter Savage Landor. Ei deulu ef oedd piau'r ystad tan y 1950au.

Tyfodd Llwyn Celyn yn fferm ffyniannus ym meddiant cyfres o denantiaid. Fel rheol meddai tenantiaid ffermydd ystad Llanthony ar brydlesau copihold eithriadol o hael, â rhentiau a ddelid am hyd pedair oes yn hytrach na'r tair oes arferol. Arhosodd yr elw o'r tir yn eiddo i'r ffermwyr, a gododd ffermdai cedyrn eraill, mawr eu bri, yn Nyffryn Ewias yn yr 16eg a'r 17eg ganrifoedd.

Ailwampiwyd Llwyn Celyn yn sylweddol yn ystod y 1690au. Ychwanegwyd llawr, simddai a grisiau yn y neuadd agored, yn ogystal â chegin wrth ochr y coridor traws.

Ond cynyddodd rhentiau'n aruthrol trwy gydol y 18fed ganrif wrth i'r hen brydlesau ildio i rai byrrach eu cyfnod a oedd yn fwy buddiol i'r tirfeddianwyr. Yn y tai allan yr arferid buddsoddi adnoddau yn Llwyn Celyn o hyn ymlaen yn hytrach na'r prif dŷ: yr ysgubor ddyrnu, y tŷ seidr, bloc y stablau, yr odynau brag a grawn, y tylcau moch ac ati. Tua 1850 fe basiodd y denantiaeth i'r teulu

Jasper, ac wedyn ym 1944 o Jack Jasper i'w frawd-yng-nghyfraith Tom Powell, a brynodd y rhydd-ddeiliadaeth a 176 o erwau ym 1958. Pasiodd hyn i lawr i'w ddau fab Trefor a Lyndon Powell ym 1990. Erbyn hynny roedd yr adeiladau wedi dirywio'n arw, ac yn 2008 gofynnodd Cadw i'r Landmark Trust ddod i mewn. Codwyd cyllid i'r perwyl, ac erbyn yr hydref 2018 roedd y gwaith o adfer Llwyn Celyn wedi dod i ben. Mae ar gael heddiw i unrhywun ei rentu ar gyfer gwyliau.

Tom Powell, who farmed at Llwyn Celyn from 1958–1990.
Tom Powell, a bu'n ffermio yn Llwyn Celyn rhwng 1958 a 1990.

After the Dissolution

In 1546, Nicholas Arnold, a rich lawyer, bought the Llanthony Priory and its Monmouthshire estates, including Llwyn Celyn, from the Crown. The Llanthony Estate remained in the ownership of the Arnolds until 1726 when it was sold to Edward Harley, whose descendants became the earls of Oxford and Mortimer. They sold the estate in 1799 to Colonel Sir Mark Wood, who sold it to the poet Walter Savage Landor in 1807. The estate remained with Landor's descendants until the 1950s.

Through all this, Llwyn Celyn was in use as a farm, and well into the 18th century, its tenants were prosperous enough to claim the status of gentlemen. Llanthony tenants enjoyed unusually generous copyhold leases, with rents held over four lives rather than the more usual three. In a period of rising food prices, the profits of the land stayed with the farmers who built the unusual number of well-constructed, high status farmhouses in the Llanthony Valley in the 16th and 17th centuries.

In the 1690s, Llwyn Celyn underwent a significant refurbishment, carried out by its tenant William Watkin and his brother Thomas. Its open hall was ceiled over to create a first floor, and a chimneystack and staircase were added, along with a kitchen off the cross passage. In the same years, the beast house was built with stalls for a dozen cattle, and plausibly the cider house too. The threshing barn predates a little building added to it in 1695/6.

The memorial to Job Watkin, Gent. in Cwmyoy Church. Cofeb i Job Watkin, Gŵr Bonheddig, yn Eglwys Cwm-iou.

The Watkins' fine high table, dated 1690. This stood in Llwyn Celyn's hall until taken in lieu of rent in the 1940s. Bwrdd godidog y teulu Watkins, sy'n dwyn y dyddiad 1690. Safai yn Llwyn Celyn nes cael ei gipio yn lle rhent yn y 1940au.

By now, it was a prosperous farm of some 150 acres, and Thomas's son Job Watkins (d. 1756) is described as a 'Gent.' on his memorial plaque in Cwmyoy Church.

But all was not as stable as this seems. The Arnolds had been locked in bitter legal disputes with their tenants since the mid-17th century, as the tenants fought to preserve their customary copyhold rights against the landlords' drive to convert them into shorter, more profitable leasehold tenancies, even though the reciprocal duties owed to a feudal lord has long since been eroded. Llwyn Celyn is a prime example: in 1717, Thomas Watkin paid a rent of just 1s 4d pa for 56 acres, apparently still under the lease for four lives that began in 1597.

The tenants eventually lost the rent war, but the Arnolds gave up the struggle too, selling the Llanthony Estate to Edward Harley in 1726. Harley's managers were efficient in bringing the estate management up to date, raising rents considerably. By 1775, Llwyn Celyn and its acreage was leased for a rent of £50 pa, a telling illustration of rising rents in the 18th century. From this point on, Llwyn Celyn, while always a significant farm in the valley, began a long decline in prosperity.

There were no more major refurbishments of the farmhouse. Instead, resources were put into its stone outbuildings – stable block, malt and grain kilns, piggery and so on. In the early 1850s, the tenancy passed to the Jasper family, who were well-known local figures, and in 1944, from Jack Jasper to his brother-in-law Tom Powell.

In 1958, Tom and his wife Olive bought the farm from the Landor Estate. For the first time in its history, Llwyn Celyn was owned by its inhabitants. Tom

The upper hall today. The floor was inserted in the 1690s. Llawr uchaf y neuadd heddiw. Gosodwyd y llawr newydd yn y 1690au.

Powell too was a local figure, and annual sheep sales were held in the field below Llwyn Celyn. At his death in 1990, his two sons, Trefor & Lyndon, inherited the farm of some 40 acres. By now, the buildings were in serious decline and Cadw erected an emergency roof over the farmhouse. Eventually, in 2008, Cadw contacted the Landmark Trust and asked if the charity would step in to give this ancient house a secure future. Landmark does this by restoring buildings at risk and offering them to everyone for holidays, and so the long process of fundraising and restoration began, completed in autumn 2018.

Pennu dyddiad Llwyn Celyn
Dating Llwyn Celyn

Mae ffrâm bren Llwyn Celyn wedi cael ei dyddio trwy gyfrwng techneg arloesol a ddatblygwyd ym Mhrifysgol Abertawe ar gyfer Prosiect Derw'r DU. Trwy'r dechneg hon, gellir sefydlu ym mha flwyddyn y disgynwyd y coed a ddefnyddiwyd trwy ddadansoddi'r isotopau ocsigen sydd ynddynt. Llwyn Celyn yw'r adeilad cyntaf erioed i gael ei ddyddio trwy'r dechneg newydd, a dilyswyd y canlyniad trwy ddadansoddiad Carbon-14.

Gweddnewidiwyd ein gwybodaeth am y safle gan y canlyniad hwn, sy'n profi bod y prif dŷ wedi cael ei adeiladu ar un cynnig ym 1420, yn gynharach o lawer na'r hyn a dybid gynt. Dangosodd ail gam yr ymchwiliad i nenfwd y neuadd a'r simnai gael eu hychwanegu tua 1690, a bod y gegin gefn wedi cael ei chodi tua'r un adeg, ynghyd â'r beudy. Mae'n debygol bod y tŷ seidr hefyd yn dyddio o gyfnod 1690, a bod yr ysgubor ddyrnu wedi cael ei hadeiladu rhwng 1656 a 1690.

Welsh timber can be difficult to date by tree ring analysis. Growing conditions are so constant that growth is too constant for the differing ring widths needed to identify a given year of construction, and until late in the restoration project, Llwyn Celyn's timbers yielded no date fix. Then Landmark was put in touch with the UK Oak Project, a climate research programme at Swansea University

One of the timbers to be sampled was the mantel beam in the inserted chimneystack.
Un o'r preniau a samplwyd oedd y trawst yn y simnai a ychwanegwyd.

Some of the timber cores used to date Llwyn Celyn. Rhai o'r creiddiau pren a ddefnyddiwyd i ddyddio Llwyn Celyn.

in association with the Oxford Dendrochronology Laboratory. This analyses the signature of the oxygen isotopes in the timber to identify its felling year. Landmark funded parallel Carbon-14 analysis as validation of the experimental results for Llwyn Celyn, which became the first previously undated building to be dated using this new technique.

The results transformed our understanding, revealing a clear felling date of 1418/19 for timber in the roof of solar range, and 1420/21 for a joist in the service end. This confirmed that the main house was built in a single phase, and much earlier than expected.

A second phase of research then proved that the hall ceiling and chimneystack were inserted *c.*1690, with the rear kitchen added around the same time. The beast house was added in the same years, and probably the threshing barn and cider house too. William and Thomas Watkin were tenants at the time, and were adding to Llwyn Celyn's land holdings, making it entirely plausible that they were the authors of this refurbishment, as Llwyn Celyn reached its peak. The date is quite late for such adaptations of the traditional medieval living spaces, but change comes slowly to such remote areas.

Drilling out a timber core in the rear kitchen. Drilio craidd pren allan yn y gegin gefn.

Disgrifiad o Lwyn Celyn
Description of Llwyn Celyn

Llwyn Celyn ym 1420: tŷ neuadd agored

Adeiladwyd Llwyn Celyn ym 1420 ar ffurf neuadd agored ac iddi dair cilfach, coridor traws ac adran gwasanaeth ar ddwy lawr. Nodwedd dra anghyffredin yw'r heulfa ddeulawr, yn cynnwys ystafelloedd preifat a ddefnyddid gan bennaeth y teulu. Ar ben uchaf y neuadd mae mainc sefydlog yr arglwydd i'w gweld o hyd, a daethpwyd o hyd i olion y canopi neu nenlen a ymestynnai unwaith uwchben yr esgynlawr lle'r eisteddai wrth ei fwrdd.

(Mesurai bwrdd uchel Llwyn Celyn dair troedfedd ar ddeg o hyd; bu yn y tŷ tan y 1950au. Rywbryd yn ystod y degawd hwnnw, aeth asiant Landor â'r bwrdd yn lle rhent. Cafodd ei werthu gan ddisgynyddion hwnnw tra oedd Llwyn Celyn yn cael ei adfer, am bris – gwaetha'r modd – na allai Landmark mo'i fforddio.)

Yn wreiddiol roedd gan y neuadd gwpl to addurnol ac ynddo agoriad yn ei ganol. Rhoddai hyn olwg ar y ddau ddrws ar draws y coridor sgrîn. Mae'n bosibl bod y gorddrysau cymhleth hyn yn unigryw mewn tŷ preswyl yng Nghymru. Arweiniai'r drysau hyn at stordai. Gellid ddringo ysgol i gyrraedd yr ystafell uwchben. Mae'n debyg mai allanfa troethle neu bisdy oedd y sianel garreg sy'n ymwthio o'r wal allanol.

Arweiniai gorddrws braf arall, ac arno ddwy darian, at adain yr heulfa. Ni wyddys a oedd y tarianau hyn wedi'u paentio. Yn ystod y gwaith adfer daethpwyd o hyd i agoriad carreg caeëdig yn y neuadd; tybir mai p'un ai drws oedd yma, yn arwain at risiau tro y tu mewn i'r muriau a esgynnai i'r siambr odidog ar lawr cyntaf yr heulfa, neu gilfach storio ac arddangos.

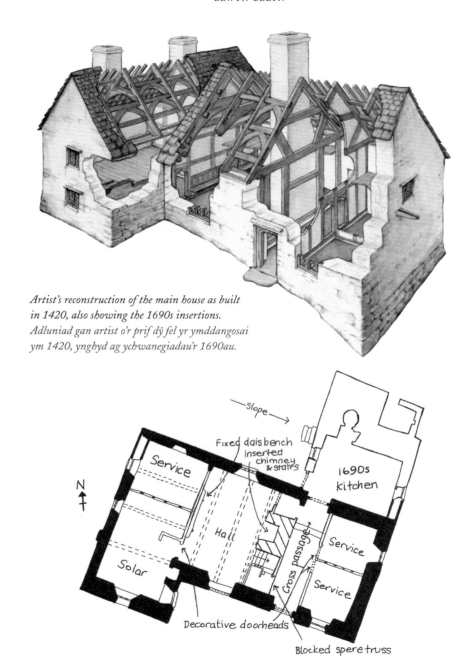

Artist's reconstruction of the main house as built in 1420, also showing the 1690s insertions.
Adluniad gan artist o'r prif dŷ fel yr ymddangosai ym 1420, ynghyd ag ychwanegiadau'r 1690au.

Llwyn Celyn in 1420: an open hall house

As first built in 1420, Llwyn Celyn was a three-bay open hall, oriented down the hillside. It had a typical cross (or screen) passage and a two-storey service end, of two rooms on the ground floor and a single chamber above. This counts as a large house for the period in Wales.

The hall was open to its roof timbers, to display the fine arched wind braces. An open hearth was in the middle of the hall, with smoke filtering up to escape through the roof, perhaps through a louvered opening. At the high end of the hall, the lord's fixed bench survives, an exceptional survival. Traces were found of a canopy that once stretched above the dais, or raised platform, on which he sat at his table.

The ceiling of the solar bedroom was probably inserted in the 1690s.
Mae'n debygol i ystafell wely'r heulfa gael ei hadeiladu yn y 1690au.

The cross passage, with the 1420 screen partition and spere truss now filled by the inserted 1690s chimney stack to the right. Y coridor traws, gyda phared sgrin a chwpl to gwreiddiol 1420, a ildiodd le yn y 1690au i'r simnai ar y llaw dde.

Leading off the high end of the hall is an additional two-storey solar cross range, private rooms for the lord. This is another highly unusual feature in a such a house, especially as it was built at the same time as the rest.

At the low end of the hall, some of the original plank-and-muntin screen survives to the right (*see p.16*) of the fireplace, as does a small stretch of the dentellated cornice that once ran all round the hall.

The hall was built with a spere truss, a decorative roof support with a central opening. Here, this gave a view of the two doors of the service range across the screen passage. The wooden frame of this opening, which is now filled by the chimney stack, can still be seen in the cross passage. At Llwyn Celyn the spere truss opening played its part in creating a ceremonial space designed to impress, since the two doorheads in the passage have decorative, ogee-shaped wooden doorheads. Their spandrels are carved with elaborate tracery such as is more usually found in a church, and in stone, and they are possibly unique in a domestic context in Wales.

These doors led to a buttery (for storing liquid provisions like wine and beer) and a pantry (for dry goods, from the Latin *panis*, meaning bread). The room above the buttery and pantry was probably reached by a simple ladder. A stone spout protrudes from its outside wall that led into a stone bowl, found hacked off behind later internal plaster. This was probably a urinal.

The ground floor of the solar range was reached through a door with another fine, ogee-shaped doorhead, this time with two bevelled shields (*see p.25*). Whether they were painted, and with what, is all part of the mystery of who built Llwyn Celyn, and why. This doorway certainly marked a place of privacy and status.

During the restoration work, a blocked arched stone doorway was discovered under later plaster at the high end of the hall in its southern wall. This may have been an intramural spiral staircase leading to the fine chamber on the first floor of the solar block, whose south facing window (also re-discovered during restoration) offers a fine view of Bryn Awr. Alternatively it may have been an alcove for storage and display.

This shoe, possibly 17th-century, was found under the eaves. Shoes were often hidden to ward off evil spirits. It was left it in situ. Dan fargod y tŷ y daethpwyd o hyd i'r esgid hon, a wnaed efallai yn yr 17eg ganrif. Arferid cuddio sgidiau fel hyn yn aml i gadw ysbrydion drwg draw. Penderfynwyd ei gadael lle roedd hi.

The 1690s chimney stack thrusting up through fine roof timbers of the once open hall (during restoration).
Simnai'r 1690au'n ymwthio i fyny drwy breniau to godidog y neuadd, a oedd gynt yn agored hyd y to
(yn ystod y gwaith adfer).

Llwyn Celyn tua 1700: adferiad

Tua diwedd yr 17eg ganrif, fe gyflawnwyd gwaith adfer a moderneiddio sylweddol yn Llwyn Celyn, a hynny mwy na thebyg gan William a Thomas Watkin. Mae'r newidiadau hyn yn nodweddiadol o ddatblygiad tai preswyl, a'r cyfnod hwnnw a adlewyrchir yn y modd a adferwyd y tŷ gan Landmark.

Gosodwyd nenfwd yn y neuadd agored i greu llawr uchaf, â simnai anferth ar hyd llinell y coridor traws, gan gau agoriad y cwpl to. Ychwanegwyd grisiau pren, gan ailgylchu rhannau o sgrîn a chornis y 1420au, yn ogystal â'r paneli uwchben y fainc sefydlog yn y neuadd. Crëwyd y ffenestri dormer ar flaen a chefn y llawr cyntaf, ynghyd â'r gegin ddeulawr fawr â'i ffwrn bara yng nghefn y coridor traws. Adeiladwyd y beudy (yr ystafell hysbysrwydd bellach) tua'r un adeg. Mae'n debyg mai'r teulu Watkins a gododd yr ysgubor ddyrnu hefyd, a'r tŷ seidr cadarn yng nghefn y ffermdy.

Llwyn Celyn *c.*1700: refurbishment

In the late 17th-century, William Watkin and his brother Thomas, prosperous farmers benefiting from long leases with low rents, carried out a major refurbishment and modernisation of Llwyn Celyn. The changes they made are typical of the evolution of the domestic house, and it is this phase that Landmark's restoration presents.

The Watkins inserted a ceiling into the open hall to create a first floor, with a massive fireplace and chimneystack along the line of the screen to the

The wooden doorhead to the solar holds two shields, now blank. Mae gorddrws pren yr heulfa'n cynnwys dwy darian, sydd heddiw yn wag.

After restoration, the hall looks much as it did c.1700 when the ceiling and chimneystack were inserted.
Yn sgil y gwaith adfer mae golwg y neuadd yn dra thebyg i'r hyn ydoedd tua 1700 ar ól i'r nenfwd
a'r simnai gael eu mewnosod.

cross passage, blocking the opening in the spere truss. They inserted a wooden dogleg staircase, initially at the left of the fireplace, but soon after moved to its right as we see today, recycling some of the 1420s screen and cornice to create it. They also installed the panelling above the fixed bench in the hall. The intramural stair or alcove in the hall, now redundant, was probably blocked up at this date. The first floor dormer windows front and back were created.

The Watkins also built the large, two storey kitchen block with bread oven leading off the rear of the cross passage.

The beast house (today's information room) was built around the same time, and it is plausible that the threshing barn, wide opposing doors aligned to catch the breeze across the threshing floor, and the well-built, cider house at the back of the farmhouse also date from the Watkins' time. Fragments of finger moulding were found on plastered ceiling joists in the upper solar room. These probably also date to c.1700 (now fully reinstated), as do the doors to the upstairs solar rooms.

Llwyn Celyn ar ôl 1800: fferm ffyniannus

Ychydig o newidiadau a wnaed i'r tŷ ar ôl canol y 18fed ganrif. Mae'r rhan fwyaf ohonynt wedi cael eu dadwneud yn unol â'r weledigaeth o ran atgyweirio a gytunwyd gan Landmark wrth adfer y safle, sef dychwelyd at y cyfnod o gwmpas 1700 (h.y., yn anad dim, blocio drws ar law dde'r fainc sefydlog a osodwyd i greu mynediad at adain yr heulfa; dileu coridorau a pharwydydd ar y llawr cyntaf, ac adfer y wal rhwng y pantri a'r bwtri).

Ymhlith y nodweddion sydd wedi goroesi o ddiwedd y 19eg ganrif a dechrau'r 20fed yw'r ffenestri Crittal â'u dolenni nodweddiadol ar ffurf dyrnau ynghau, a ddefnyddiwyd yn helaeth ar ystad Landor.

Roedd yna gytiau moch hefyd; odyn i sychu brag i facsu cwrw ac un arall ar gyfer gwenith; stablau; perllan; gardd lysiau a chwt ieir. Ond erbyn blynyddoedd olaf yr 20fed ganrif, roedd hyn i gyd wedi mynd â'i ben iddo.

The kitchen was added to the rear of the house in the 1690s. Ychwanegwyd y gegin ar gefn y tŷ yn y 1690au.

Llwyn Celyn from 1800: a thriving farm

Few changes were made to the house after the mid-18th century as rents rose. Those that were made survived into living memory, but most have most been reversed in accordance with the agreed philosophy of repair for Landmark's restoration to of a return to *c.*1700 (chiefly the blocking of a door inserted to the right of the fixed bench; the removal of a wooden porch, and corridors and partitions on the first floor, and the reinstatement of the wall dividing pantry and buttery).

Later features that do remain from the late-19th or early-20th century are the Crittal windows with their distinctive clenched fist handles, widely used in Landor estate properties.

By now the farm was an extensive, self-sufficient unit, long-existent activities formalised into well-constructed stone outbuildings typical of farmsteads in

By 1800, Llwyn Celyn had become an integrated unit of specialised outbuildings. The cider house is to the right (today a comfortable bedroom and bathroom). Erbyn 1800 roedd Llwyn Celyn wedi tyfu'n uned integredig yn cynnwys tai allan arbenigol. Saif y tŷ seidr (sydd bellach yn ystafell wely ac ystafell ymolchi) ar y dde.

the Black Mountains. Llwyn Celyn thrived, combining cattle, sheep and arable farming. There was a two-storey piggery; a kiln to dry malt for brewing beer and another for wheat; stables; an orchard; a kitchen garden, and a hen coop. Cows were milked in the beast house, and the cool, dark room to the right of the fixed bench became a dairy for making cheese and salting meat. But by the closing years of the 20th century, this ancient form of life had dwindled and the paternal oversight of a landed estate no longer existed as a safety net to ensure upkeep of the farm's buildings.

The front door to Llwyn Celyn in 2007.
Drws ffrynt Llwyn Celyn yn 2007.

The stairs at Llwyn Celyn,
inserted in the late 1600s.
Grisiau Llwyn Celyn, a
ychwanegwyd ar ddiwedd y 1600au.

These clenched fist window latches were used in Llanthony
estate farms in the early 20th century. Arferid gosod y
dolenni ffenestr hyn ar ffurf dyrnau ynghau yn ffermdai
ystad Llanthony ym mlynyddoedd cynnar yr 20fed ganrif.

Adfer Llwyn Celyn

Roedd prosiect adfer Llwyn Celyn yn gryn fenter, a ofynnai am sgiliau llu o weithwyr proffesiynol a chrefftwyr. Mae crefftwyr wedi cael profiad gwerthfawr wrth ddysgu technegau traddodiadol: defnyddiwyd to teils carreg Llwyn Celyn yn arbennig fel cyfrwng i drosglwyddo dulliau traddodiadol o gynhyrchu a gosod teils a oedd wedi cael eu cloddio mewn chwarel yng Nghwm Olchon gerllaw. Aethpwyd ati mewn ysbryd ceidwadol i gyflawni'r gwaith atgyweirio saernïol, a gellir gwerthfawrogi'r canlyniad wrth ochr y preniau canoloesol. Defnyddiwyd morteri a phlastrau calch traddodiadol yn ddieithriad.

Mae nifer o'r hen dai allan yn cael eu defnyddio'n ysgafn erbyn hyn gan y gymuned neu'r cyhoedd. Mae Llwyn Celyn yn cymryd ei le bellach yn yr 21ain ganrif, yn dra ymwybodol o'i orffennol tra'n edrych ymlaen at ddyfodol diogel yng ngofal Landmark.

Llwyn Celyn's walls have been invisibly tied back together using injected resin ties. Llwyn Celyn yn erbyn cefnlun ei dirwedd; mae'r patrwm caeau hynafol i'w weld o hyd.

Restoration of Llwyn Celyn

The restoration of Llwyn Celyn was a huge undertaking, involving the skills of many professionals and craftsmen. The walls were invisibly tied back together using resin ties. The ancient timber frame was straightened, patiently inching dropped trusses and doorways back into their former positions. Younger craftsmen have gained experience in traditional techniques: Llwyn Celyn's stone-tiled roof especially was used to transmit traditional methods of tile production and laying, the tiles quarried in the adjacent Olchon valley. Extensive joinery repairs were made conservatively, and the necessary

Opposite: Llwyn Celyn in its landscape after restoration, ancien field patterns still apparent. Cyferbyn: Llwyn Celyn yn erbyn cefnlun ei dirwedd; mae'r patrwm caeau hynafol i'w weld o hyd

repairs can be appreciated alongside the medieval timber. Lime mortars and plasters were used throughout. Most of the techniques and materials used would be recognised by the late medieval men who built the house just after Glyn Dŵr's demise, but modern advances of course find their place, in underfloor heating and the installation of bathrooms in previously unusable spaces.

Several of the former outbuildings are now in low-key community or public use, reflecting the sort of popular leisure activities unthinkable in the harsher times of previous centuries, but perhaps reinventing the significant role once played by this site commanding the valley entrance. Llwyn Celyn now takes its place in the 21st century, entirely mindful of its past while looking forward to a secure future in Landmark's care.

Left: Straightening the dropped arch brace in the hall roof.
Chwith: Sythu'r trawsbren gollyngedig yn nho'r neuadd.

*Below: Skilful timber repairs were needed to save as much as
possible of the original fabric. Isod: Bu angen gwaith atgyweirio
medrus i achub cymaint â phosibl o'r adeiledd gwreiddiol.*

*Left: Setting out a
traditional Welsh
valley detail on the
cider house roof with
massive new stone
tiles. Chwith: Gosod
cafn traddodiadol
Cymreig ar do'r tŷ
seidr, gan ddefnyddio
teils cerrig anferth.*

Llwyn Celyn, in its prime location looking across to the Skirrid, just as it has for the last 600 years.
Llwyn Celyn, yn edrych draw tuag at Ysgyryd Fawr ers 600 o flynyddoedd.

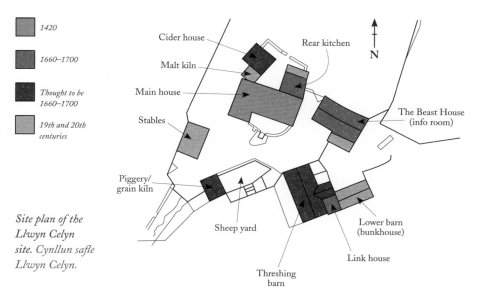

1420

1660–1700

Thought to be
1660–1700

19th and 20th
centuries

Cider house

Rear kitchen

N

Malt kiln

Main house

The Beast House
(info room)

Stables

Piggery/
grain kiln

Site plan of the
Llwyn Celyn
site. Cynllun safle
Llwyn Celyn.

Sheep yard

Lower barn
(bunkhouse)

Link house

Threshing
barn

The beast house (now an information room).
Tŷ'r anifeiliaid, sydd bellach yn ystafell
hysbysrwydd.

The threshing barn, built in the mid 1600s.
Yr ysgubor ddyrnu, a adeiladwyd tua chanol
y 1600au.